GRAVITY &

THE UNIFIED

FIELD

THEORY

Greenhill http://greenhillspublishing.com.au/

Walsh,

Martin (author)

GRAVITY & THE UNIFIED FIELD THEORY

ISBN 978-1-923156-40-1 (paperback)

PHYSICS

Cover by Green Hill Publishing

GRAVITY &

THE UNIFIED

FIELD

THEORY

MARTIN WALSH

CONTENTS

PREFACE

With thanks firstly to God the Father, Jesus and the leading and guiding of the Holy Spirit with all facets contributing to the piece and my life. Mum, Dad, Michelle, Ashley, Shanna, Brooke, Madison, Cain, Janice and David, Dani Kelly with specific contributions, Creflo, Joyce Meyer, Joseph Prince, David Jeremiah, Michael Youssef, Socrates, Plato, Euclid, Kepler, Copernicus, Galileo, Descartes, Maxwell, Pascal, Noether, Newton, Leibniz, Johann Bernoulli, Faraday, Planck, Bohr, Oppenheimer, Bose, Einstein, Dirac, Pauli, Feynman, Hawking, The Alpinist and Marc-Andre Leclerc, Tommy Caldwell, Kevin Jorgeson, Alex Honnold, Daniel & Sheria W, Wayne M, David Stirling, Jock Lewes and their original party, Ben Macintyre, Ed and Steve Sabol and NFL Films.

The concept of quantum gravity proposed in this work is observed and evident in the natural world. Self-evident truths that can be seen and repeated in clear observations are stated as to remove all grounds of doubt. Strictly adhering to the conservation of energy law, Noether's theorem, Newton's 3 laws of motion, Pauli exclusion principle restrictions, Bose-Einstein statistics, Photoelectric effect, standard model of particle physics, 9.81ms2, KE=mv2 and the literal kinetic energy in objects in the field of energy from the earth's gravitational field of energy.

In the spirit of presenting this piece with much simplicity, to involve and explain to the greatest possible audience all effort has been attempted. Feynman, "explain it to me as if I were a 5-year-old child", Einstein, "if you cannot explain it simply enough you do not know it well enough". Saying that, the first piece is understanding space and time and is complex and vital. Paradoxical and symmetrical events take place and themselves can take some effort to grasp, as well is the removal of preconceived notions. There are some functions that require large sentences with several commas, this is necessary to gain the whole concept by viewing all facets together. Rene Descartes, "The first thing was never to accept anything for true, which I did not clearly know to be as such, that is to say; carefully to avoid precipitancy and prejudice and to comprise nothing more in my own judgement than was presented to my mind so clearly and distinctly as to remove all grounds of doubt", EQ.

The conservation of energy law has been noted as far back as Thales of Miletus (550 B.C) and Empedocles (490-430 B.C). The conservation of energy law states that the total energy of an isolated system remains constant, it is said to be conserved over time. Energy is neither created nor destroyed, rather it can only be transferred or transformed from one form to another. For instance, chemical energy is converted into kinetic energy when a stick of dynamite explodes. If one adds up all the forms of energy that were released in the explosion, such as the kinetic energy and potential energy of all the pieces, as well as heat and sound, one will get the exact decrease of chemical energy in the combustion of the dynamite.

1.

MANIFOLD OF

SPACE TIME

Working from "the construct of time, a forward time frame" described by Stephen Hawking and all contributors of his team at the time, so massively clever people. "Into the universe with Stephen Hawking-the theory of everything", and "A brief history of time".

"Before the big bang the fog of energy had nowhere to exist, so viewing it from the outside was impossible". End quote, EQ.

Fusing space and time into four dimensions, VS three dimensions and time. Definition, "In physics, spacetime is any mathematical model that fuses the 3 dimensions of space and the 1 dimension of time into a single four-dimensional continuum", EQ.

Working from the big bang of Hawking: Before the big bang we see there was no space and there was no time, no space, no time of space exiting constantly over time. Then space started to exist (after the big bang) and exist constantly and continually, we have the spacetime continuum. The structure of space exists constantly and that constant existence can be timed over time. Now the time tied to space, spacetime, is as constant as the constant existence of space itself. Time itself/the aspect of time, is directly associated with the on-going existence of space. Because space exists constantly it exists over time, and this is where we get time and the aspect of time from. Space and time are stated as 1 word, spacetime. To be united into a 1-word description both space and time obviously share an equality, "things which are equal to the same thing are equal to each other", Euclid. The time of space and that space existing over time is constant. Next is to actions around the big bang/renovation of energy.

There was nowhere to expand into before the big bang.

There was no on-going action to transform something into more space, it was the big bang and not the big banging.

There was no item to transform into more space nor was it necessary, the finalized item was achieved with the big bang. An action with the result of a light year always being a light year. The James Webb telescope shows no edge of space and as a result we experience no up and down of/in space. This is absolutely the reason why we do not experience up and down in space.

There was an item of energy (loose term fog), a single action (big bang) to a finalized result, space and the fermions in it, the universe. Space goes from nowhere to everywhere as a single action to result. The Minkowski spacetime metric, Euclidean space and parameters in the 2nd postulate of Special Relativity (SR) all show space as a constant with a light year always being a light year's distance of space. Whilst the individual components in Euclidean space and Minkowski spacetime may vary due to length contraction and time dilation, all frames of reference (Euclid, Mink) will agree on the total distance in spacetime between events.

MANIFOLDS: Euclidean space, and, Minkowski spacetime.

Euclid of Alexandria, fourth century B.C. is referred to as the father of geometry. From the Euclidean and Minkowski propositions are 2 items, space and time, and geometry and time, with all being constant.

2 propositions that need to be addressed are: 1, "expanding space" (Hubble) and 2, "local spacetime curvature" (General relativity, GR).

1, Quote Hawking, "for approximately 330,000 years after the big bang the energy cooled to its current state". EQ. During this time space is already everywhere and no stars have yet been formed. The first stars then took approximately 500 million years to form. After all this time, once a star or galaxy is formed space does not just start to move the star or galaxy. Newton's 1st law of motion shows objects in space do not interact with space, so not only is space not expanding, space does not interact with objects in it moving the objects, that would not adhere to Noether's theorem either. The last piece of evidence is the Andromeda galaxy still sits where it is relative to us, and shows it must be travelling at "exactly" the same speed, and in "exactly" the same direction/the identical velocity, or space is simply not moving everything apart.

2, The primary core concept of GR is "local spacetime curvature", Wheeler "Matter tells space how to curve, and curved space tell matter how to move". Quote Albert Einstein, "As a result of the more careful study of the electromagnetic phenomenon, we have come to regard action at a distance as a process impossible without the intervention of an intermediary medium". EQ. For space to be curved due mass, it is vital there is interaction between space and mass leading to the alteration/curvature of space.

- Newton's 1st law of motion states "An object will remain at rest or in uniform motion in a straight line, unless compelled to change its state by an external force". Observing space is never the "external force" due to space already offering rest and uniform motion in a straight line, it cannot now offer alteration from those it already offers. Space does not offer alteration from "rest" due to non-interaction, and space does not offer alteration from "uniform motion in a straight line" due to non-interaction.

- NASA designed a booster that is used to move and stop in space that works off/pushes off itself to promote motion and stoppage due to non-interaction with space. There is literally nothing to push off to promote motion or stoppage, no interaction. This does not state that space is nothing, simply

  a characteristic of space is non-interaction with mass and most energy formats in it.

- Conserved momentum whilst travelling in space shows space does not interact with an object in motion, showing no interaction means no alteration.

- The 2nd postulate of SR shows space does not interact with light, hence light travelling in a straight line. Again, there is no alteration of light and no interaction means no alteration.

There are 4 literal physics examples of non-interaction between space and mass, and as seen with electromagnetism, if there is no interaction there is no alteration. Adhering to physics evidence it is physically impossible for space to be curved due to mass in it. With physics evidence we can clearly observe space is not expanding and there is no "local spacetime curvature" proposed by GR.

Space offers a space with complete freedom of motion through noninteraction, so fermions can act and interact through their intrinsic characteristics generating kinetic energy through motion. Kinetic energy is then transferred (Newton's cradle) or transformed per fermion, hadron, atom or molecule. Space is also cold and most energy formats transform into heat through 1 or more process which is then absorbed into the cold of space. This is an action of space and its characteristics, working in tandem with the fermions and their created energy formats in space regulating heat creation, all from the one item before the big bang. This action also shows us the four fundamental interactions will unify due to all being of/from one energy format.

Now we have a relevant understanding to characteristics of space and time and its tandem interactions with energy formats in it, and with clear physics evidence and perspective that space does not curve due to objects in it, the next observation is our earth's gravitational field of energy. Observing the actions of gravity right in front of us and the literal kinetic energy in objects in the earth's gravitational field from the field of energy we live in.

2

GRAVITATIONAL

FIELDS

INTRODUCTION

"Energy show the velocity of mass, and mass shows the velocity of energy".

Quote, "Mathematically, momentum (mv) is identical to kinetic energy (mv2), apart from the squared. In Newtonian mechanics, momentum (momenta or momentums; more specifically linear momentum or translational momentum) is the product of the mass and velocity of an object. It is a vector quantity, possessing a magnitude and a direction: The objects momentum is p=mv, p (push, drive), m (mass) and v (velocity)", EQ. The original kinetic energy formula created by Leibniz and J. Bernoulli is K.E.=mv2, this is the correct formula relative to a correct amount of kinetic energy in acceleration. The ½ in the kinetic energy formula K.E.=1/2mv2 is ONLY from a "Taylor expansion series" of infinite numbers, and was added to the kinetic energy formula due to SR and large mass numbers. Providing smaller numbers to work with, but an inaccurate and incorrect amount of kinetic energy expression per the conservation of energy law and Noether's theorem. *FOOTNOTE expansion of K.E.=1/2mv2 at the end of the works.

Standing on the shoulders of giants, William of Ockham (1287-1347) and Hawking (1942-2018).

Johannes Kepler (1571-1630), "Nature prefers simplicity over complexity, if she can find a simply way of doing something, she will unhesitatingly choose that way rather than something more complex", EQ.

Rene Descartes (1596-1650), "All truths were linked to one another, so that finding a fundamental truth and proceeding with logic would open the way to all science". EQ.

Actions of kinetic energy (including pendulums) have been observed by Galileo (1564-1642), Christiaan Huygens (1629-1695), Gottfried Wilhelm Leibniz (1646-1716), Johann Bernoulli (1667-1748), Emilie du Chatelet (1706-1749) and Sir Isaac Newton (1642-1727), among others.

Observing the earth's gravitational field of energy Galileo found at sea level objects fell at 9.81ms2. From this, working with the simplest of terms to explain all aspects only using examples drawn from our physical universe. In adherence with Newtonian physics which is discussing only aspects of interactions that can be seen and repeated. Since we are looking for physical evidence to show reality in these works, physical examples are the foundation, in this way the natural world validates the piece. All examples are self-evident truths for the reader to simply look in front of you and conclude for yourself, gravity is right in front of you.

The earth's gravitational field is a field of kinetic energy operating as a fundamental interaction, this means the action and energy format cannot be reducible to more basic/fundamental interactions. "Energy shows the velocity of mass,": The size and strength (fall rate, 9.81ms2) of the field is determined by the speed of rotation, composition of the mass and parameters of kinetic energy in a rotational pattern. The 2 parameters of energy inputs and mass determines the size of the field. "Mass shows the velocity of energy": The mass with a central axis creates a warped field of kinetic energy, the warping produces correction and the correction provides velocity information of the field to the centre, 9.81ms2 at sea level. Kinetic energy is transferred from the field to objects in the field with direct contact transference. Transference stated in the conservation of energy law and seen in Newton's cradle, earth's field of energy and the blades of a wind farm turbine and a moving atmosphere, so the wind. Kinetic energy can exhibit motion and relativistic weight/mass. When an object sits on your table it has "potential" energy in it, this is the kinetic energy from the field exhibiting relativistic weight/mass, observed as potential energy. When the object falls from your table the potential energy does not transform (we do not observe that) into kinetic energy, it always was kinetic energy. The transformation aspect (from the conservation of energy law) in this action relates to transforming from relative rest on the table, into motion moving to the floor.

Observing that kinetic energy is motion and relativistic weight/mass, to go twice as fast you need four times the amount of input or kinetic energy, the K.E. input goes into the motion and relativistic weight/mass.

The mass and central axis creates a warped field of kinetic energy which is under constant corrective parameters through "characteristics of kinetic energy with warping" (chapter 10). The constant corrective and motion in space creates the acceleration constant.

Quote Albert Einstein, "The body produces a field in its immediate neighbourhood directly; the intensity and direction of the field at points farther removed from the body are thence determined by the laws which govern the properties in space of the gravitational field themselves".

3.

FOUR FUNDAMENTAL

FORCES OR

INTERACTIONS

Quote scientific literature: The fundamental interactions, also known as the fundamental forces, are the interactions that do not appear to be reducible to more basic interactions. There are four interactions known to exist: the gravitational and electromagnetic interactions which produce significant long-range forces whose effects can be seen in everyday life, and the strong and weak interactions, which produce forces at miniscule subatomic distances and govern nuclear interactions. Each of the known fundamental interactions can be described mathematically as a field. The strong interaction is carried by a particle called the "gluon" and is responsible for "quarks" binding together to form "hadrons" such as "protons" and "neutrons". As a residual effect, it creates the nuclear force that binds the later particles to form atomic nuclei. The weak interaction is carried by "W and Z bosons" and also acts on the nucleus of the atom mediating radioactive decay. The electromagnetic force carries the photon and creates electric and magnetic fields which are responsible for the attraction between orbital electrons and atomic nuclei which holds atoms together, as well as chemical bonding and electromagnetic waves, including visible light. Although the electromagnetic force is far stronger than gravity it tends to cancel itself out within large objects. EQ.

To change lines of thought with relevance, the "theory of everything" or T.O.E. Quote, "The theory of everything is a single, all encompassing, coherent theoretical framework of physics that fully explains and links all aspects of the universe. Two theories which all modern physics rests upon are General Relativity (GR) and Quantum field theory (QFT). GR is a theoretical framework that only focuses on gravity for understanding the universe in regions of both large-scale high mass, galaxies and clusters of galaxies etc. On the other hand, QFT has successfully implemented the standard model of particle physics that describe three non-gravitational forces: strong, weak and electromagnetic force. Physicists have experimentally confirmed virtually every prediction made by these two theories when in their domain of applicability. Nevertheless, GR and QFT are mutually incompatible – they cannot both be right. Since the usual domains of applicability are so different, most situations require only one of the two are used. The compatibility between GR and QFT is only an issue in regions of extremely small scale, the Plank scale – as those that exist within a black hole or during the beginning of the universe. To resolve this incompatibility between GR and QFT, a deeper underlying reality, unifying gravity with the other three interactions, must be discovered to harmoniously integrate the realms of GR and QFT into a seamless whole. The theory of everything is a single theory that in principle is capable of describing all phenomenon in the universe. According to theory the four fundamental forces were once a single fundamental force".

Quantum gravity: Difficulties arise when applying the usual prescription of QFT to the force of gravity via bosons, at this present time called "graviton Bosons". There is no complete quantum field theory of gravitons due to an outstanding mathematical problem with renormalization in GR, in theory gravity is a massless state of a fundamental string.

4.

EFFECTS OF GRAVITY

PHYSICALLY SHOWN

ON THE EARTH

Physical examples that gravity displays on the earth will give a solid foundation to expand into the following chapters. Building upon physical examples presented with Newtonian physics, working from the literal kinetic energy produced from the earth's gravitational field, and adhering strictly to Newton's 3 laws of motion, the conservation of energy law and Noether's theorem. Newton's 3 laws of motion and the conservation of energy law work perfectly together, validate each other and accurately describe actions and aspects of kinetic energy, verifiable through the natural world around us.

1, Objects in the earth's gravitational field experience a literal result of kinetic energy. This is taking place from the earth's gravitational field of energy and direct contact to the objects in the field through the action of transference of kinetic energy.

2, The field of gravitational energy has velocity parameters to the centre of the field. At sea level to the rate Galileo found and Sir Isaac Newton used, 9.81ms2.

3, Gravitational energy has increase at a constant as the earth rotates at a constant.

4, The size and strength to the field of energy is determined by the speed of rotation, composition of the mass, the mass with a central axis and parameters of kinetic energy in a rotational pattern.

5, Defying a gravitational field is different than removal of its energy fingerprint.

6, Albert E. "The body produces a field in its immediate neighbourhood directly, the intensity and direction of the field at points farther removed from the body are thence determined by the laws which governs the properties in space of the gravitational fields themselves".

7, Albert E. "As a result of the more careful study of the electromagnetic phenomenon, we have come to regard action at a distance as a process impossible without the intervention of an intermediary medium".

8, " ….. GR enables us to derive theoretically the influence of a gravitational field on the course of natural process". EQ.

9, Objects in the earth's gravitational field are given relativistic weight/mass (potential), motion and relativistic mass, and are taxed of their intrinsic energy provided by an alternate source to the gravitational field. In the case of potential energy, kinetic energy always needs to be accounted for and it will exhibit relativistic weight/mass (potential) only, or relativistic mass with motion.

"Energy shows the velocity of mass, and mass shows the velocity of energy".

5

GRAVITATIONAL ENERGY SHOWN AS KINETIC ENERGY

All objects in the earth's gravitational field display literal kinetic energy, and through the actions described in conservation of energy law and Noether's theorem it is either transformed or transferred. We can observe the literal kinetic energy of the field when we observe the action of mass and a weighing scale. When an object is placed on the scale the amount of mass and the relativistic mass with velocity information of the field is displayed as relativistic weight in the earth's gravitational field. Taking the scale into space away from the gravitational field of energy the mass does not register on the scale and simply sits there next to the scale. The relativistic weight with velocity information and the mass gives us the scale determination of weight and field. Now, attaching the scale to a large heavy block that simply sits there still in space. If we threw a marble at the scale, the kinetic energy on impact would give a 1-time value register of the kinetic energy to the relativistic mass and the mass. This demonstrating the kinetic energy from the earth's gravitational field through a weighing scale. All objects in the earth's field have kinetic energy from the field either as motion and relativistic mass or just the relativistic mass, that people knew as "potential energy".

6.

ALBERT EINSTEIN,

OBSERVATION OF A

MAN IN A CARRIAGE

Albert E. stated, "Let us imagine ourselves transferred to our old friend the railway carriage, which is travelling at a uniform rate. As long as it is moving uniformly, the occupant is not sensible of its motion, and it is for this reason that he can, without reluctance, interpret the facts of the case indicating that the carriage is at rest and the embankment in motion". EQ.

If the atmosphere was not moving with us in the carriage, Einstein's man in a carriage would be man on a carriage. From this, once in motion at a constant we do not notice the kinetic energy in us or around us. The atmosphere within the carriage is also moving with the occupant and carriage, so, to its physical capabilities it would also house a relative amount of kinetic energy, simply for the fact of it being in motion.

A wind farm turbine has the kinetic energy of a moving atmosphere, wind, transferred to the blades which is then transformed through a generator into electricity.

7.

PHYSICAL MECHANICS

OF THE EARTH'S

GRAVITATIONAL FIELD

Adhering to the conservation of energy law and Noether's theorem, "energy shows the velocity of the mass, and the mass shows the velocity of energy". We have direct parameters of input equals output, relative to size and strength of the field adhering to Newton's 3rd law of motion and Noether's theorem". The 1st law of motion shows information of rest (no energy) and motion (evidence of kinetic energy) and non-interaction parameters to characteristics of space. Also, it shows us the "first/underlying field dynamics" (first law of motion/energy) of the underlying format in an electromagnetic field, kinetic energy. A linear path described in the 1st law of motion creates an even "body of energy"/field that resides within the body due to the linear motion. This is an even field of distributed energy through a body and in a body when created, and acting through a linear path VS a rotating one.

The mass with a central axis is a rotational pattern that creates a warped field of kinetic energy. The INPUTS are the speed of rotation, parameters of kinetic energy in a rotational pattern creating correction, and the composition of the mass. The OUTPUTS are the size of the field and the rate in which the field corrects itself to the centre, at sea level 9.81ms2. Space with non-interaction conserves momentum, the conserved rotational momentum produces constant warping of the field that generates the correction constant. As a result, this creates the acceleration constant with a mass in free fall and characteristics of kinetic energy.

8.

SOLAR SYSTEM

The solar system field and the evidence it provides also brings validation to the earth's gravitational field of energy, as it should. When it comes to space you are always in it but you never touch it. Nicolaus Copernicus (1473-1543) stated that the planets operated in this manner.

Column A = Relative distance from the sun

Column B = Time to go once around the sun

	A	B
Mercury	.4	88 days
Venus	.7	225 days
Earth	1	365 days
Mars	1.5	687 days
Jupiter	5.2	11.9 years
Saturn	9.5	29.5 years

Actions of the earth's gravitational field, the solar system, Orion's spur and the Milky way galaxy all have identical results of motion due to the identical energy format being used, in space. To put a link in a historical chain:

1, The sun revolves around the earth.

2, The earth revolves around the sun.

3, Our galaxy and its action of rotation has formed due to the interaction of two black holes. The one in the centre of our galaxy passing through and destructing the other leaving our black hole and kinetic energy and all manner of fermions, atoms that later produce heavy elements of our Milky way galaxy.

To observe our solar system accurately, it is vital to see from the centre of the Milky way galaxy, Orion's arm and our solar system, starting from the current black hole in the centre of our galaxy. Working from chapter 11. "Considering we now know all actions happen in space and not to space. A star is comprised of atoms that are made up of fermions and energy formats of the atomic orbitals. Before a star collapses you could travel all around it observing it sits in space and not on space. The collapse pressure takes places to the atoms in space and not to space. The collapse forces the hadrons and fermions right next to each other unable to form atoms, with the fermions STILL obeying the Pauli exclusion principle restrictions. Due to the small size of a fermion, it will only ever take on physical pressure relative to its size. Being extremely small outwards pressure can only manifest per its physical size. The collapse forces the fermions and hadrons next to each other, essentially a massive nucleus including electrons, and with massive energetic potential and properties. This makes a black hole the densest object we can see in space. If the collapse is even to the centre the black hole (sphere of particles) will not spin, as with the black hole in the centre of our galaxy. If the collapse is uneven to the centre, slightly to the left, the sphere will spin to the magnitude of the collapse. The actions and parameters of earth's gravitational field show the size and strength parameters to a gravitational field. The density of the sphere and the magnitude of the collapse through to rotation, creates a gravitational field like no other we see in space, as we observe with spinning black holes".

The black hole of our Milky way galaxy came into contact with a larger one. The interaction demolishing the larger black hole spreading galactic arms of kinetic energy spiralling, with fermions, hadrons, atoms and elements of our galaxy. We have observed gravitational waves emitted from the interaction of 2 black holes, which is also the identical energy format of "the fabric", (kinetic energy) to the spiral arms of our galaxy. With and at the start of our solar system there is a certain speed in relation to the beginning of our solar system disc. This is approximately the area where Neptune resides. Our solar system is positioned at the rear of Orion's spur. Due to being at the rear of the spur a vortex is created with the fabric of the spur through to the rotation of the galaxy. The vortex dynamics circles into increasingly smaller circles, increasing in speed to the centre through vortex dynamics. Once the fabric meets at the centre, the vortex inward spiral dynamics stops and circular orbits remain. The fabric item of our solar system still being attached to Orion's spur bringing it around with the spur of the spiral galaxy. The suns ingredients hydrogen and helium are the smallest and lightest atoms and molecules, and are the only ones that could have physically travelled at the beginning on the fabric to the centre of the solar system disc. Other heavier elements not being as light did not reach the centre and remained out in the disc. The sun is trapped in the middle of the solar system disc due to lighter elements and formation action of the solar system disc.

The physics explanation to the actions with the fabric item of our galaxy and our solar system starts with the following chapter, "the example". In strict adherence and describing actions of the "BoseEinstein statistics", (B.E.s) Satyendra Nath Bose (1894-1974). From the B.E.s the mechanics will be made physically evident. The subject of the solar system and kinetic energy expressions form one subject whilst intermittingly dominating the subject matter. Mercury and its eccentric orbit will be explained at the relevant place of mention with the correct equation to its orbit.

9.

THE EXAMPLE

From this example a physical process will be followed with Newtonian physics.

The example: First, an object travelling in space free from any gravitational field or galactic influence. The object is travelling in a straight line (axis x) and spiralling right to left or left to right (axis Z), a tumbling end over end action can be added, the results do not change. Operating on two axis X, Z as described in Euclidean geometry: Newton's 1st law of motion states, "An object will remain at rest or in uniform motion in a straight line unless compelled to change its state by an external force". Uniform motion to both of the inputs promoting the motion over both X and Z axes. In free space this action continues unless altered by an external force with the 2 inputs of kinetic energy existing in the single body. Completely unhindered due to each other, no interaction or combination and complete overlap of the 2 inputs of kinetic energy. This action behaves identically to the characteristics of the B.E.s with the energy format stated in the conservation of energy law and functions described by Newton's 3 laws of motion, kinetic energy. The Bose-Einstein statistics usage will be made evident.

Bose-Einstein statistics: "In quantum statistics, Bose-Einstein statistics (B.E.s) describes one of two possible ways in which a collection of non-interacting, indistinguishable particles may occupy a set of available energy states at thermo-dynamic equilibrium. The aggregation of particles in the same state, which is a characteristic of particles obeying the B.E.s accounts for cohesive streaming of laser light and frictionless creeping of "superfluid helium". The B.E.s applies only to particles not limited to single occupancy of the same state – that is, particles that do not obey the Pauli exclusion principle restrictions. Such particles are named bosons". End quote.

From here we will observe individual aspects of the example. The object will be put into motion with boosters in relevant areas. The booster will function and deliver a quantum of energy that is transformed into kinetic energy and residing within the body in a quantifiable amount. Once each potential energy has transformed and the kinetic energy is acting upon the body through the 2 axes, the bodies/field of kinetic energy are present and quantifiable, responsible for the motion and overlapping each other within the single body. There is an overlap of kinetic energy in the same body (state) without hindrance, interaction, combination or taxing of the two. This action, relative to the energy format kinetic energy, is observed and described with the example and a photon (boson). Kinetic energy is stated with transform and transfer, and is used in multiple actions of energy expressions. This B.E.s subject continued through to chapter 10 and 11.

To see the action of the planets travelling on the fabric of the solar system disc with the overlap parameters of the B.E.s in mind, we will now "turn that inside out" and see it relative to the planets gravitational fields and solar system disc.

Dealing with the fabric (K.E.) of the disc to our solar system and the mechanics of the planets on it. The solar system disc of kinetic energy is responsible for the "carpet ride effect" to the planets residing on/in it with the sun trapped in the centre. The planets have their own intrinsic energy moving them around the disc resulting in the altering speeds of the planets. Actions such as Shoemaker-Levy 9 take place as are unforeseen influences contributing to this. The disc imparts the action of circular orbits, the kinetic energy associated with the orbital actions of the planets is from the solar system disc of kinetic energy. It is imparted on a planetary body and does not affect the rotational speed or gravitational forces produced by the planets. Focus will be on the body (planet) and where it is, not the gravitational energy associated with the body. Movement in our solar system is either given by the disc or it is external from the solar system, apart from the disc and associated with the object. The disc is imparted on the planets with an example being: the solar system disc is like a carpet and the planets are on the carpet. When the carpet moves all move together, and the planet is somewhat not moving relative to the carpet. With the start to the action of the solar system there is a certain speed in relation to the beginning of the solar system disc, roughly the area where Neptune resides. Our solar system and its positioning create the relevant vortex dynamics to the disc of our solar system action with increased orbital speed approaching the centre.

The solar system disc is reflected (with relative applicability) in the speeds to some of the planets, Neptune and Mercury are 2 of these, also is the altered speeds of the planets during Perihelion and Aphelion. Speeds to some planets reflects possible action of occurrence leading to alteration such as Shoemaker-Levy 9. The disc providing gravitational effects (circular orbits) of our solar system gives Mercury its ability to escape the "so called" gravitational effects of the sun. Warping to the solar system disc – the effects to the warping (to whatever extent) takes place to a planetary gravitational field 1 and the fabric of the solar system disc 2, with the following results: A satellite is governed by its internal energy promoting its motion, the gravitational field working on the body (satellite), the solar system disc which is tied to the action of Orion's spur, the freedom of motion space provides (1st law of motion) and the B.E.s action to the energy involved. The action of motion to a satellite around earth brings various parameters from different origins on a single object in a specific area, each exhibiting the existence of their presence, working in adherence to physics laws and the overall function of the satellite. This area of interaction is complex, to a relevant extent involves the moon and extends to a relevant distance with corresponding results.

Consideration: this does not state the sun has no gravity, but for the sun to have its own gravity pulling the planets around it, then Orion's spur it is associated with must have its own force to bring the sun around with Orion's spur. For this to be the case, why does this force not have direct effect on the planets relative to the sun, considering it has enough strength to influence the sun, but not planetary motion of the solar system?

The numbers to this piece have come about due to the study of the presented works, the numbers speak for themselves. The only accuracy decision is with the individual numbers and their relevance to be in the equation. Their deniability or acceptance I attempt to address successfully. Per person, I am confident that the applicability of usage rest upon each individual and their acceptance. However, the actions and numbers physically remain physically evident and realistically exists, things which can be seen and repeated. The numbers presented are: The mean speed of Neptune, its position from the centre of the solar system and its equality to other objects in the solar system. The mean speed of Mercury, its position from the centre of the solar system and its equality to other objects in the solar system. The solar system disc has a place and an object (planet Neptune) that we can derive a speed from, and is close to the position that is the beginning of the solar system.

(Always makes me think of Reuben Tishkoff, "the guy, the place and the thing") Planets and items are noted within this equation, taking note of all throughout and their representation to each other. This area has an asteroid belt between Mars and Jupiter, and it to need representation in the equation. Things which are equal to the same thing are equal to each other, Euclid. If we regard a planet with a numerical value of 1 and classify those in the group with equality, the asteroid belt is not equal to the planets and their value of 1. The chosen number selected can be found and validated numerically through the literal physical actions presented with physics. Eight planets plus an asteroid belt, speeds of Neptune and Mercury and the position of start and finish (inner and outer) of solar system. The speeds offered are available from: nssdc.nasa.gov.

Mean speed of Neptune = 5.43 km/s, planet 8.

Mean speed of Mercury = 47.36 km/s, planet 1.

8 planets.

Asteroid belt value = .72191529.

Planets plus belt = 8.72191529.

5.43 x 8.72191529 = 47.36.

Mean speed of Neptune.

Planets plus belt.

Mean speed of Mercury.

KINETIC ENERY

CHARACTERISTICS

Gottfried Leibniz and Johann Bernoulli invented the name kinetic energy, and stated it as the living force "Vis viva". – Kinetic energy in physics must be accounted for due to it being a literal energy format described in the conservation of energy law and Noether's theorem, with the formula for kinetic energy in acceleration K. $E=mv2$. Newton's 1st law of motion also provides "field dynamics" to the underlying unifying field of electromagnetism, the "original unifying field dynamics". The 1st law of energy/motion.

Starting from Newtons 1st law of motion and observing an object at rest and then in uniform motion in a straight line. This action is taking place in an area of space free from any external gravitational force. The object is a rectangle and its mass throughout is considered consistent. The object will move from rest (point 1) through to constant motion (point 2). Areas of the body are noted as A = front, B = centre, C = rear. Motion is an action and energy format created/generated in a space that offers complete freedom of motion, which is then transferred or transformed per fermion, hadron, atom, molecule or orb in Newton's cradle. Viewing the energy body of kinetic energy within the body that is literally in existence, massless and quantifiable. The areas of the body; front, centre and rear travel the identical distance and the field of energy is then even throughout the body showing an "even field dynamic". Kinetic energy is "benign apart from motion", the motion aspect (velocity) is a representation to the quantifiable amount of kinetic energy, as 1 electron volt is a representation to a quantifiable amount of electricity. In a natural process of uniform motion in a straight line a field of kinetic energy within the body is "even", "massless", "quantifiable" and "benign apart from motion". The next pattern of motion is a rotational pattern generating a warped field of kinetic energy. Through creation of an uneven field, and with the action of correction by the field we observe an "elastic" characteristic to kinetic energy. The inputs to a rotational pattern determine the size of the field producing a field of kinetic energy then utilized as a gravitational field of energy. Kinetic energy is the "cleanest energy format" and is used in gravitational actions (actions of motion), with absolute intrusion, without any significant alteration on a cellular, molecular or atomic level, with no leftover waste product regardless of an action. Able to manifest on a fermion, hadron, atom, molecule or orb in Newton's cradle. In Newton's cradle kinetic energy "transfers" through the individual orbs on contact. Kinetic energy is "benign" and its field can be overwritten with electricity and magnetism and is the energy format that propagates light at the maximum speed of kinetic energy.

The overlap aspect to atomic orbitals during covalent bonding operates through the B.E.s showing atomic orbitals are comprised of kinetic energy, electrons and any electromagnetism relative to the nucleus and electron interaction. Electrons, energy from electron and nuclei interaction and kinetic energy, transforms the kinetic energy into the relativistic mass of an atom. This is observed and validated through the gold foil experiment and the exhibited relativistic mass of kinetic energy. Atomic orbitals produce chemical energy which is an expression of kinetic energy including the action of transforming into heat. Kinetic energy characteristics include:

Being an action and simultaneously a quantifiable usable energy format.

Benign apart from motion.

Can be overwritten with electricity and magnetism.

Is the propagation energy format to an electromagnetic wave, light.

The underlying field dynamic with an evenly distributed field.

Massless.

Quantifiable.

Transferable.

Can transform into relativistic mass/weight, and heat.

With varying amounts of electrons can manifest into gases and solids and exhibits characteristics of the Pauli exclusion principle restrictions.

Overlap ability stated in the B.E.s.

Elastic.

Shares an equality with fermions.

Kinetic energy has 13 characteristics.

11.

THE UNIFIED FIELD

APPLICATION AND

QUANTUM GRAVITY

"In physics, mass-energy equivalence is the relationship between mass and energy in a system's rest frame, where the two quantities differ only by a multiplicative constant and the unit of measurement. The principle is described by the physicist Albert Einstein's formula $E=mc^2$. The equivalence principle implies that when mass is lost (relativistic mass/weight) in chemical or nuclear reactions, a corresponding amount of energy will be released as radiant energy, such as light or thermal energy. The principle is fundamental to many fields of physics, including nuclear and particle physics. Rest mass also called invariant mass, is a physical property that is independent of momentum, even at extreme speed approaching the speed of light. Its value is the same in all inertial frames of reference. Massless particles such as photons have zero-invariant mass, but massless free particles have both momentum and energy". Kinetic energy and photons both have zero rest mass. Albert Einstein's mass-energy equivalence describes the relationship between mass and energy, a fermion and a field/wave and their common ground. Where the two are equal and share an equality, their equivalence. From a mathematical stand point their common denominator property of both. $E=mc^2$ NEVER SUGGESTS a fermion will transform into a field or a field will transform into a fermion. Since the only mass (m) that can travel at the speed of light (c), the only mass that will qualify for m in $E=mc^2$ is a photon.

"With the action of an induced fission reaction, a neutron is absorbed by a uranium 235 nuclei, turning it briefly into an excited uranium 236 nucleus, with the excitation energy provided by the kinetic energy of the neutron plus the forces that bind the neutron".

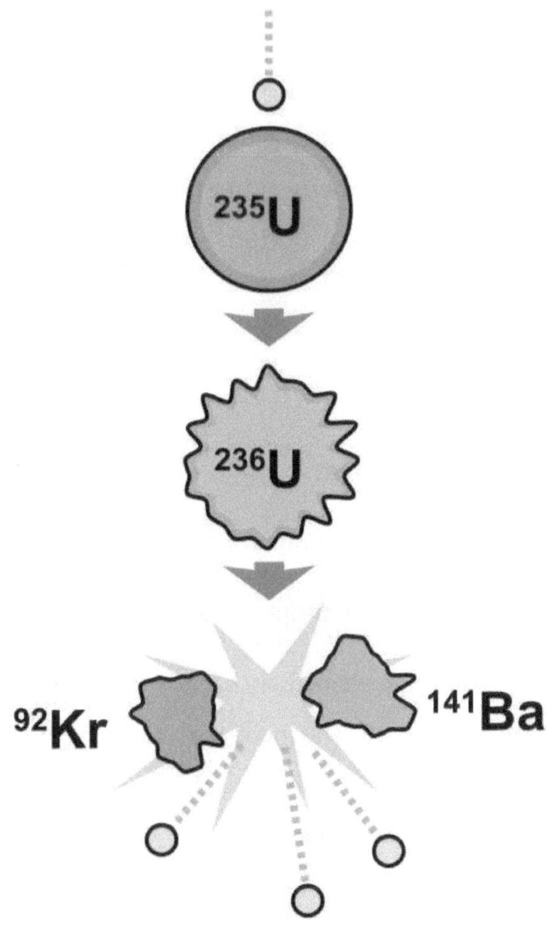

Induced fission reaction. A neutron is absorbed by a uranium-235 nucleus, turning it briefly into an excited uranium-236 nucleus, with the excitation energy provided by the kinetic energy of the neutron plus the forces that bind the neutron. The uranium-

Starting here we have invariant mass (fermions), gluons/strong interaction and kinetic energy. Kinetic energy has the ability to transfer to a fermion and a neutron, and/or have its field overwritten with the intrinsic characteristics electricity and magnetism, resulting in the excited state. From a physical perspective of evident energy formats with uranium 235, we have a neutron, nuclei and kinetic energy. From the kinetic energy formula K. E=mv2 we can define kinetic energy as E and a fermion as m with an equivalence characteristic to both. The expression with the = represents an equality, things which are equal to the same thing are equal to each other.

First observing a common ground between all fermions. All fermions have an electrical body that is neither created nor destroyed, with all exhibiting a magnetic property ACCEPT the neutrino.

NEUTRINO.

COMPOSITION: Elementary particle.

STATISTICS: Fermionic.

FAMILY: Leptons, antileptons.

GENERATION: First (Ve), second (Vu), third (Vt).

TYPES: 3 types: electron neutrino, muon neutrino and tau neutrino.

Quote: A neutrino is an elementary particle that interacts via the weak interaction and gravity. The neutrino is so named because it is "electrically neutral" and because its rest mass is so small it was long thought to be zero. End quote.

The property common in all fermions is their physical energy body described in the Pauli exclusion principle restrictions, and that is observed with the neutrino being "electrically neutral". Their energetic expression is electric, whilst simultaneously being neutral. Here we have the "electrically neutral" parameter.

Kinetic energy is benign apart from motion, its energetic expression is motion, whilst simultaneously being benign. Here we have the "benign apart from motion" parameter.

The equivalent parameter of "neutral" AND "benign" are the aspect utilized for energy transition (transference, transformation) to manifest as a field or an energized fermion or hadron.

SUMMARY: Energy-mass equivalence describes a common aspect of both fermion and field. The quantum mechanics action between an invariant (never created and never destroyed) fermion and the primary energy format created through motion in space (kinetic energy) by those fermions, who's intrinsic characteristics promote attraction and repulsion. The fermions and kinetic energy interact and create relativistic mass observed with atomic structures. Jesus is made unto wisdom for us.

The neutral/benign position exists between the two poles, + and -, of a magnet and interacts with both. The + and − are "different manifestations of the same phenomenon" (Explaining electricity and magnetism, Maxwell 1831-1879) and both exist at the furthest point from each other whilst still being together. This action is observed in an electromagnetic wave with (quote) "electricity and magnetism are different manifestations of the same phenomenon", and existing in the electromagnetic wave at the furthest point from each other whilst still being together. This represents electricity and magnetism can operate together through the pole mechanism seen with the + and − of a magnet. These specific characteristics can operate through symmetry actions and energy formats electricity, magnetism, positive and negative.

Electromagnetic Wave

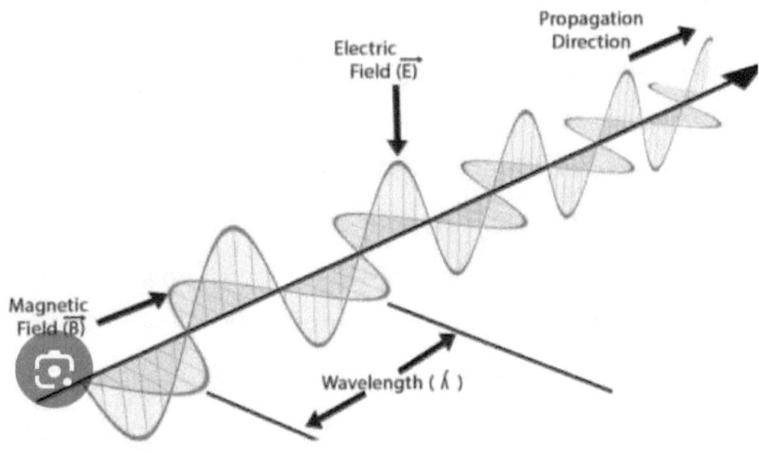

The B.E.s describes the ability of kinetic energy to occupy the same state (place in space) without hinderance, interaction or combination. This action takes place on a quantum level firstly.

1, A single object containing kinetic energy promoting motion on multiple axes.

2, The action, non-interaction and overlap of bosons.

3, Atomic orbitals overlap action during covalent bonding.

4, Planetary gravitational fields and the solar system disc.

Quantum mechanics to atomic structuring of fermions and utilized energy formats:

Fermions are simultaneously electric and magnetic with these formats present as pole mechanics, existing in a single body. Fermions being the origin of electricity and magnetism, both formats are independent features, expressed as a single item (fermion) through the characteristic of the same phenomenon with different manifestations, and different manifestations being electricity and magnetism.

An up quark has a mass of =2.2 MeV/c2, with a charge value of + 2/3, and spin of ½. The + up quark charge value represents a greater electric property. A down quark has a mass of = 4.7 MeV/c2, with a charge value of − 1/3, and spin of ½. The − down quark charge value represents a greater magnetic property.

A proton is a hadron and comprised of 2 up and 1 down quark. The quark composition is + up, - down, + up. The mechanics of a magnet is two symmetrical formats + and -, and 1 neutral format. The up quark is the + of the magnet. The down quark is the neutral of the magnet (relative to electricity). The up quark is the − (relative) of the magnet. Joined with electric, magnetic, electric (positive). Energy, neutral and energy pole formats. A proton has 2 up and 1 down quark and has a + electric characteristic.

Symmetrically, a neutron is a hadron and comprised of 2 down and 1 up quark. The quark composition is – down, + up, - down, and symmetrically has a – magnetic characteristic. A neutron is a type of electromagnet which the magnetic field is produced only by an electric current, that is generated by electrons, proton and nuclei interaction. A neutron uses aspects of magnet characteristics, but is by energy definition a monopole. High energy magnetic fields are also produced and observed with a neutron star. Here we observe that a generator with electricity and magnetism is modelled exactly on the principles of a nuclei.

An electron has a – electric representation in the standard model of particle physics, and a moving electron generates a - magnetic field.

A + electric proton attracts a - electric electron, and simultaneously the – magnetic neutron repels the – magnetic electron. The molecule aspect causes confinement of movement, and with simultaneous attraction and repulsion causing the electrons to vibrate relative to the number of electrons, giving each specific atom a specific frequency.

There are 2 electrons in the K shell with 1 allowing the path of electricity and the other magnetism, in and out of the atom through the electron path. An iron atom has a path for both electricity and magnetism through the electron configuration. A copper atom has a path for electricity through the electron configuration but does not have one for magnetism, hence copper conducts electricity but is not magnetic.

Hydrogen repels a single electron with no neutron. The down quark has an invariant mass of = 4.7 MeV/c2, and the electron has an invariant mass of = 0.511 MeV/c2. Multiplying the electron mass by 8 has a result of 4.088 which is still under the down quark mass. The – down quark of the proton repels a single electron in the hydrogen atom.

Kinetic energy exhibits motion and relativistic mass/weight, combined with electrons and a nucleus the relativistic mass we observe in atomic orbitals is present, hence we observe actions of the gold foil experiment. Relativistic mass is observed with a photon, a photon has zero-rest mass and kinetic energy has zero-rest mass. Photons are produced from atomic orbitals in a specific action and from high energy collisions. Through the photoelectric effect photons are absorbed back into the atomic orbitals, that are comprised of electrons (electricity and magnetism) and kinetic energy.

Kinetic energy produces the relativistic mass observed in a photon with a photon comprising of electricity, magnetism and kinetic energy. Gluons and W and Z bosons are comprised of identical energy formats. The large mass to W and Z bosons is through magnetism, this being the energy format that mediates radioactive decay, and is present with the neutrons used in atomic bombardment also mediating atomic splitting.

BLACK HOLES IN SPACE.

Certain stars collapse into a black hole and before these stars collapse, we could travel around them observing that the star sits in space and not on space. This adheres to Newtons 1st law of motion with the description of rest. A star is made up of atoms that are comprised of fermions and atomic orbitals. When the star collapses the pressure takes place to the atoms of the star forcing the fermions right next to each other, and still obeying the Pauli exclusion principle restrictions. The fermions right next to each other cannot create atomic structures and emit no light. The black hole is a sphere of fermions and as a result is the densest object found in space, as we have observed. If the collapse is even to the centre the black hole will not spin. If the collapse is uneven to the centre, slightly to the left, the black hole will spin to the magnitude of the collapse and generates a gargantuan gravitational field, as we also observe.

The strong and weak interaction use the properties of the fermions that is electricity and magnetism. Electricity and magnetism interact with kinetic energy forming an electromagnetic wave and gravitational actions utilizes the energy format kinetic energy. This shows the interaction between specific bosons, fermions and fields, including gravitational.

The four fundamental interactions utilize the energy formats electricity and magnetism, electromagnetism and kinetic energy.

A fractal representation is exhibited with an atoms nucleus, the electrons and orbitals comprised of kinetic energy and our solar system. The sun = nucleus, the electrons = planets, and atomic orbitals comprised of kinetic energy showing the energy format used in the gravitational actions of our solar system and earth's gravitational field of energy.

The last energy format to incorporate is heat. The physics evidence we observing relative to heat is: Water expands when it is cooled into ice. Hot water freezes faster than cold water. Iron expands with the energy format heat. Thermal expansion.

Majority of the energy format heat resides in the area between the nucleus and K shell, as a result when gases are compressed, they ultimately become cold. In removing the area responsible for heat retention, the result is the absence of heat and therefore cold. Water that is heated before freezing experiences thermal expansion resulting in slightly larger atoms. This produces less atoms to freeze resulting in hot water freezing quicker than cold. Frozen water has a reduced area between the nucleus and K shell.

This increases rigid molecular structuring (with less freedom of movement with fluid dynamics), through decreased distance between hadrons and electrons initiating geometrical configurations of particles resulting in a slightly larger volume with frozen water. Heat added to iron increases the area between K shell and nucleus resulting in thermal expansion. With enough heat added to an atom the nucleus and electrons are forced out of electromagnetic range deconstructing the atom. Chemical reactions involve the energy formats of atomic orbitals, which are comprise of kinetic energy. Kinetic energy is transformed into heat through 1 or more process and is a product of chemical reactions.

The complete unified field application including all energy formats, actions and interactions of the isolated physical system stated in the conservation of energy law, including the tandem works of both space and energy described as our universe.

*FOOTNOTE expansion: Gottfried Leibniz created the original formula e=mv2. Leibniz and Johann Bernoulli created the term "kinetic energy" and produced the kinetic energy formula K. E=mv2. Supposedly, due to large numbers produced through Special Relativity calculations with mass, the "Taylor expansion" adds the ½ to the kinetic energy formula allowing smaller numbers to be written, this changes the formula to K. E=1/2mv2. The addition of the ½ produces an incorrect and inaccurate expression to the realistic amount of kinetic energy produced. The next action that takes place is Albert Einstein removes the v from e=mv2, replacing it with a c and creates the famous equation e=mc2, and is attributed with it.

It would have been noticed that Albert replaced the v in e=mv2 (Leibniz) with a c in his equation e=mc2, simply altering the velocity parameter to the c of light. Through relativity the statement of "motion is relative" means kinetic energy is now seen as relative. General Relativity obscures the energy format kinetic energy, stating gravitational actions are the result of "local spacetime curvature", and keeping the correct quantum gravitational field energy law from being known and accepted, until the time of the unified field theory.